THE BIG BANG

AND THE HISTORY OF THE UNIVERSE

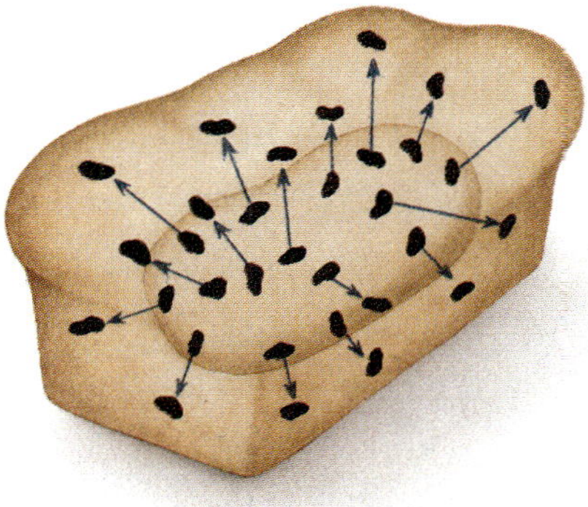

by Christina Wilsdon • illustrated by Ka Botzis

McGraw-Hill School Division
New York Farmington

Star gazing

With your family, go outside on a clear, dark night. Choose a night when the Moon's light is dim. Find a spot that is away from bright streetlights and glowing signs. Then gaze up into the sky. What will you see?

You will certainly see stars. You may glimpse just a few, if you are in a city. Or you may be dazzled by hundreds if you are in the country, far from city lights. In either case, you are seeing much more than just stars. You are also looking into the past.

How can you look into the past by gazing at stars? Stars are trillions of miles away from Earth. It takes many years for their light to travel that distance. So when you see a star, you are seeing the light it sent out many years ago. You are seeing the star as it looked in the past, not as it looks today.

This ability to see into the past occurs even when you glance at our closest star, the Sun. The Sun is 93 million miles (149.6 million km) from Earth. It takes light about eight minutes to travel that distance. So our view of the Sun is always eight minutes old. Of course, we should never look directly at the Sun because of the danger to our eyes.

Astronomers (scientists who study the universe) hope to see as far back in time as possible.

Where Did the Universe Come From?

Astronomers who study the origin of the universe are called "cosmologists." Cosmologists are working on the pieces of a puzzle called the big bang theory. A theory is an idea which explains old observations, and also predicts new observations. Scientists use the big bang theory to study and explain the history of the universe.

The big bang theory states that all matter, energy, space, and time came into being about 15 billion years ago in an event called the big bang. At the big bang the universe was unbelievably dense. For example, the matter, energy, and space that we see today were squeezed to a point smaller than the head of a pin. Ever since the big bang, space has expanded and the universe has become less and less dense. Keep in mind that the big bang did not happen at only one part of space. Rather, it happened everywhere in space. All of the space we can now see was squeezed to a point.

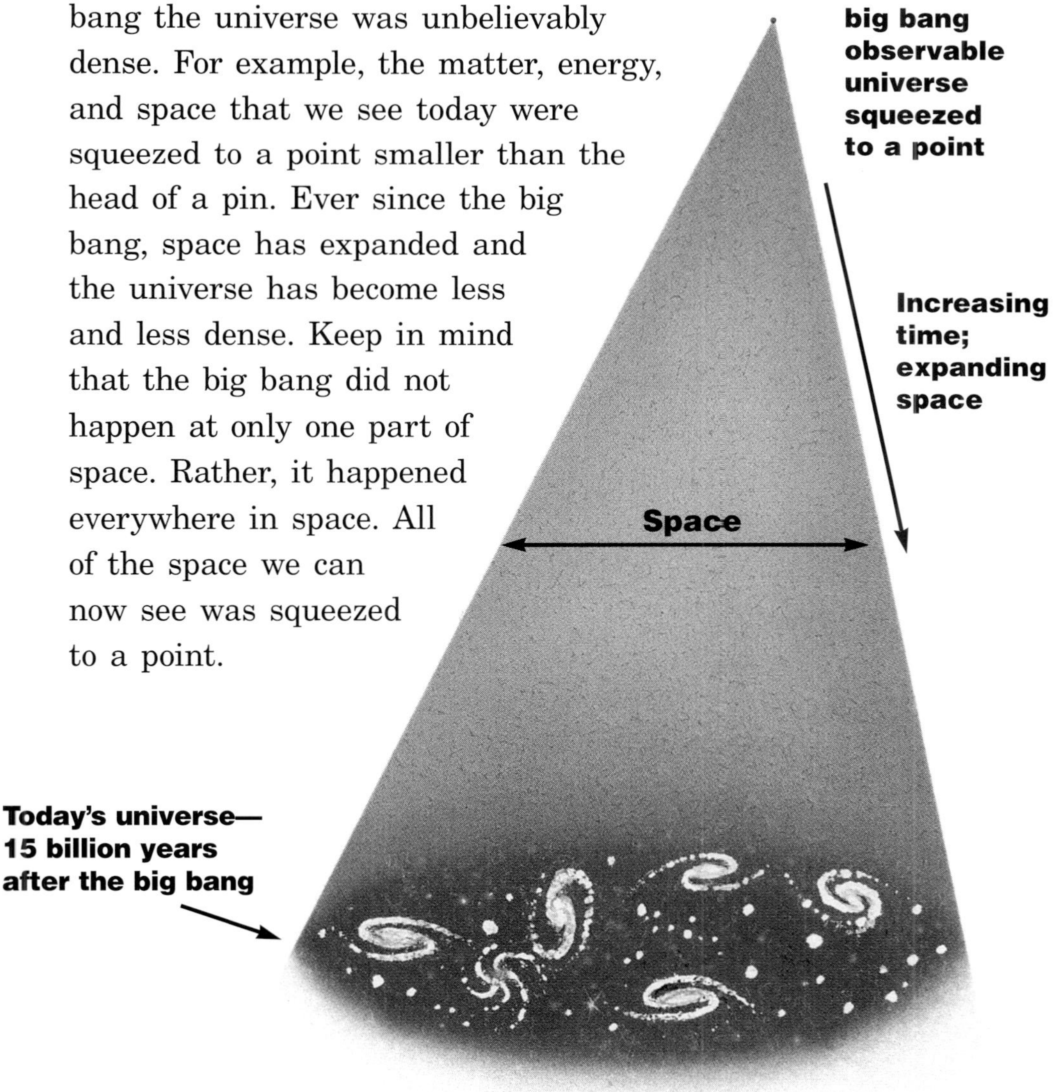

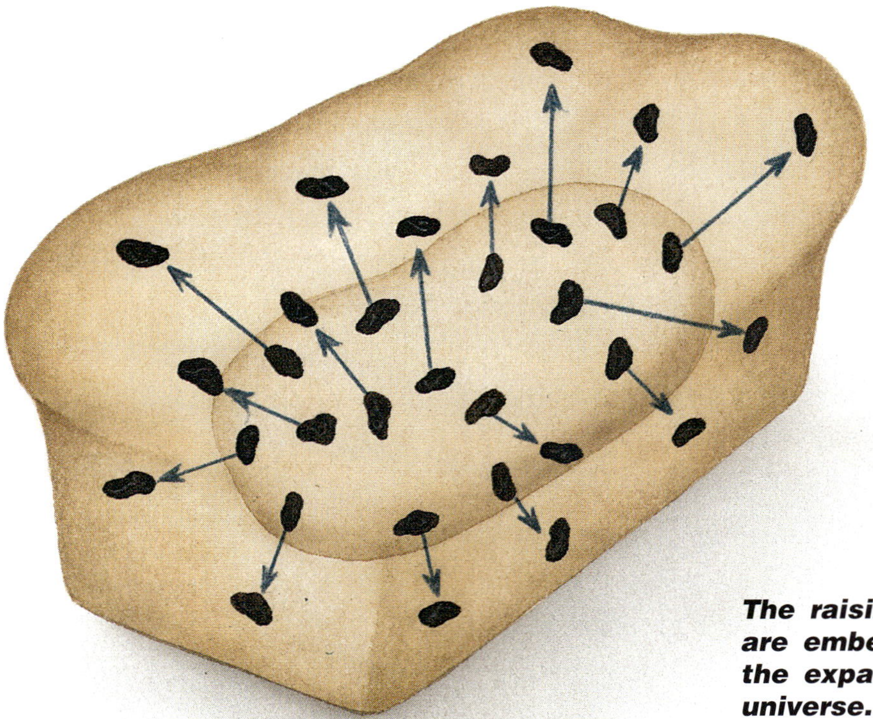

The raisin-galaxies are embedded in the expanding bread universe.

Because there is no single spot where the big bang happened, there is no single spot that forms the center of the universe. Yet the big bang is often compared to an explosion. Right after this "explosion," the universe began to grow, or expand, in all directions at the same time.

Tiny particles in the universe spread in all directions, too. But they did not fly out from a middle point like fireworks do. Instead, the particles "went along for the ride" as the universe expanded.

You can imagine how these particles spread by thinking of them as raisins in a loaf of baking bread. The raisins are close together when the bread dough is put into the oven. As it bakes, the dough rises and puffs. The dough grows larger. It carries the raisins farther and farther apart. Even though they are moving apart, the raisins remain evenly distributed throughout the dough. Just as dough expands when heated, space expands.

The early universe, however, did not grow hotter like dough being baked. Instead, it started off hot and then cooled. When the universe was one second old, it was thousands of times hotter than the center of the Sun.

The early universe was also very dense. This means that there was a very large amount of energy and matter stuffed into a small amount of space.

Energy and matter were spread out fairly evenly in the early universe, like drops of water in an ocean. But there were some bumps and ripples in this ocean of energy and matter. There were places that were a bit hotter or a bit cooler. There were places that were a little more dense than others.

These areas of greater density were the first collections of ingredients that would later become stars, galaxies, and other space objects.

HOW THE EARLY UNIVERSE CHANGED

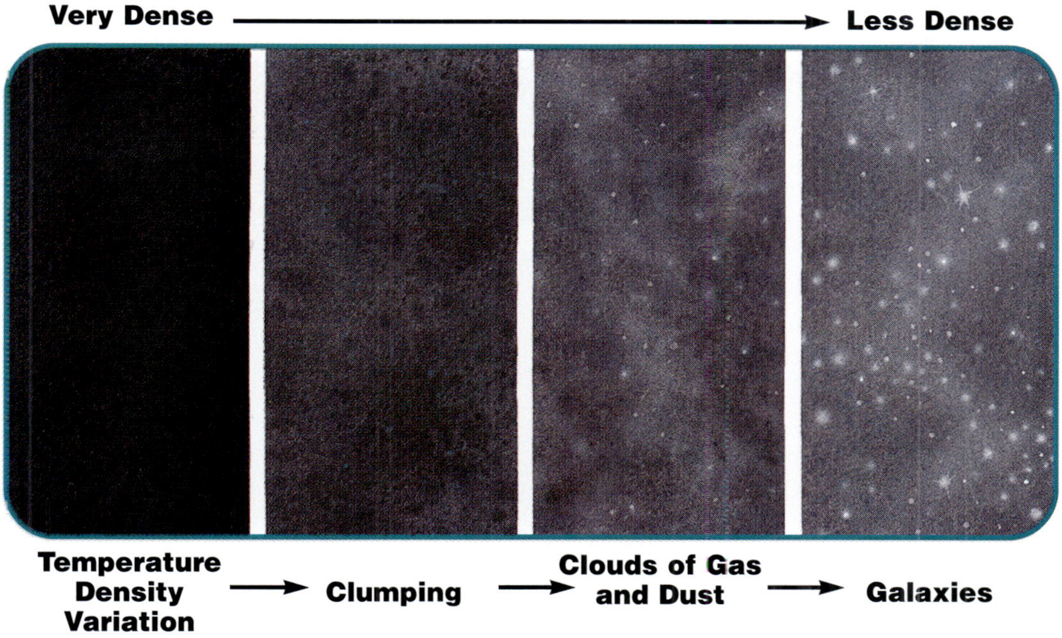

As the universe grew older, it expanded even more. As it expanded, it cooled off a little. By the time it was a few minutes old, the universe was only a few hundred times hotter than the Sun's center. The universe also became less dense as it expanded.

One hundred thousand years went by. During this time, the universe kept expanding. It also became much cooler. Its temperature dropped to just a few thousand degrees—about as hot as the Sun's surface.

What was happening to the tiny particles as the universe expanded and cooled? They were traveling through space as the universe carried them along. They were also crashing into each other. As they collided, they produced particles called protons and neutrons. By the time the universe was a few minutes old, these protons and neutrons began to combine. Together, they formed particles that we call nuclei.

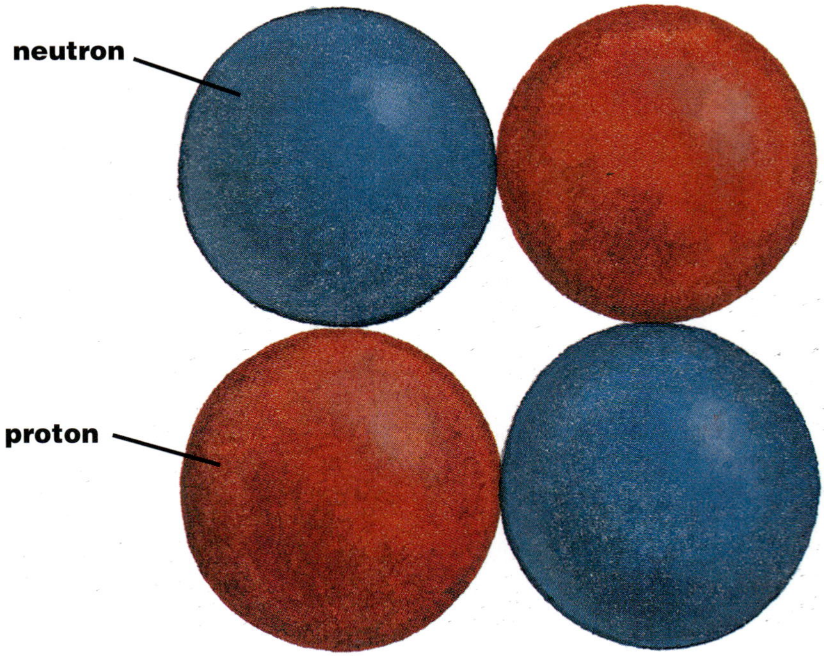

A helium nucleus, made out of 2 protons and 2 neutrons

The particles in the universe moved ever more slowly as the universe continued to cool. Some of these particles were electrons—particles with a negative electric charge. The cooling of the universe slowed these electrons. This made it possible for the nuclei to catch the electrons and combine with them. In this way, the first atoms were formed—about half a million years after the big bang.

Atoms are like tiny building blocks. They form the material that makes up gases, liquids, and solids. These early atoms formed the first elements. Elements are basic chemical substances that can't be broken down to form other substances, unless the temperature is as high as it was earlier in the history of the universe.

The first elements to form were hydrogen, helium, and lithium. Hydrogen and helium are often in the form of gases. They form the bulk of the material in stars. Lithium is a very light metal.

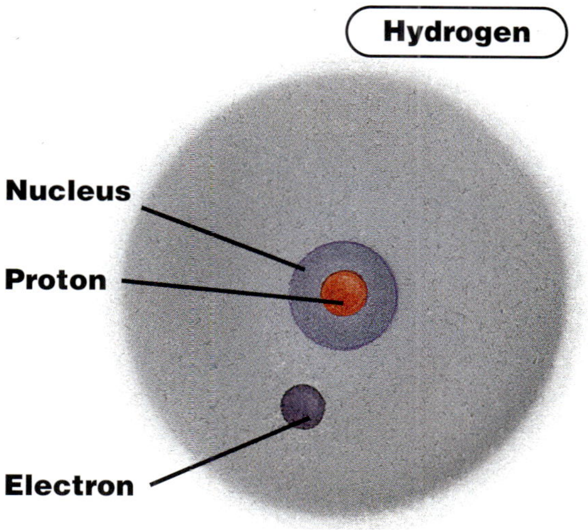

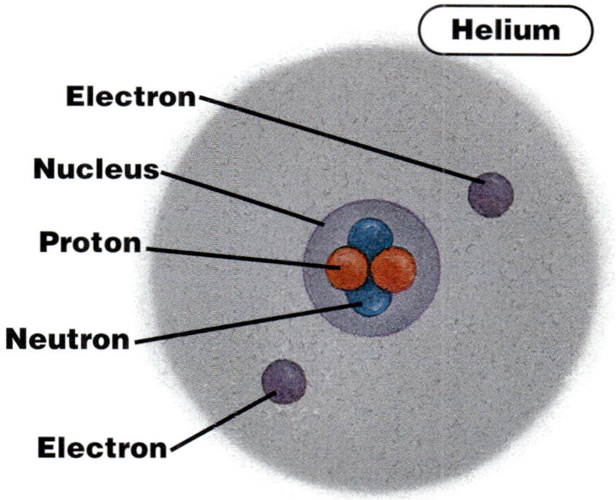

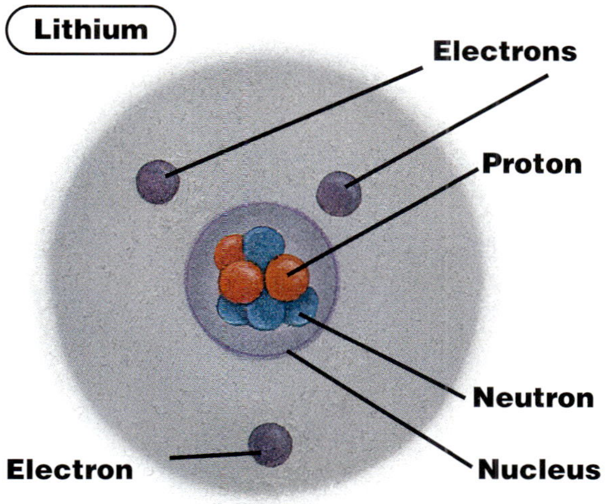

A star is born in the early universe.

As time went by, some of the matter in space began to clump together. They first began to clump in parts of the universe that were already a bit more dense than other places. These clumps attracted more matter to them. As they grew in size, their gravity became greater, too. This attracted even more matter. Gravity was forming the first clouds of gas and dust in the universe.

Layers began to form inside these clouds. Hydrogen and helium sank into the middle. Here, they were pressed beneath the weight of the gas and dust layers above them. They began to heat up under this pressure.

Finally, the balls of helium and hydrogen became so hot that the hydrogen atoms were squeezed together until they combined to form new atoms of helium. This change is called a thermonuclear reaction.

Lots of energy is given off when a thermonuclear reaction happens. It is what makes the Sun and other stars glow and give off heat. Scientists think this is how the first stars began to shine in the universe.

Inside these early stars, new elements formed. One element was oxygen—the gas we need to breathe. Another was carbon, which forms every living thing. Other elements made in the furnaces of the stars include iron and aluminum. When big stars exploded, they sent these new elements flying out across the universe. Such stars are called supernovas.

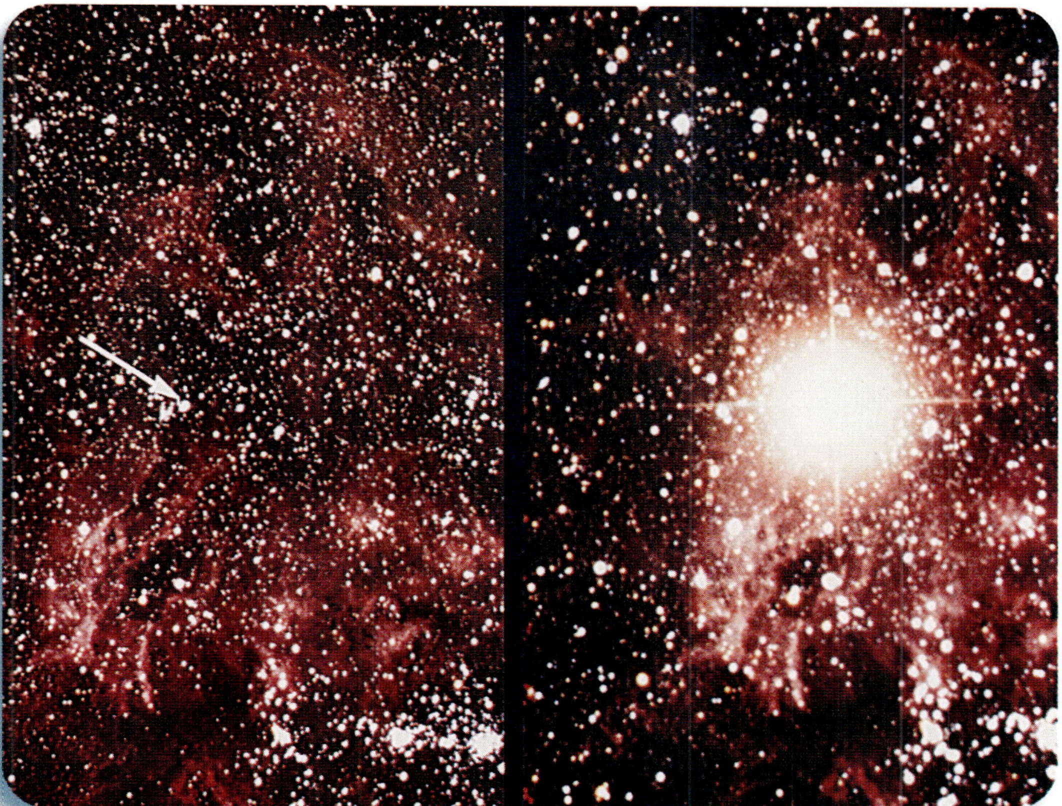

Supernova

The Milky Way Galaxy

Over billions of years, the first stars clumped together to form large groups of stars. These groups are called galaxies. New stars and planets formed after this. Our Earth and the rest of our solar system did not form until $10\frac{1}{2}$ billion years after the big bang.

Our solar system is part of a galaxy called the Milky Way. The Milky Way contains about 100 billion stars. There are hundreds of billions of galaxies in the universe. You can see some galaxies even without using a telescope.

One galaxy you can spy with your eyes is the Andromeda galaxy. Andromeda's light takes more than 2 million years to reach Earth. Scientists say that Andromeda is over 2 million "light-years" away. A light-year is the distance light can travel in a year. One light-year is about 6 trillion miles (9.5 trillion kilometers).

Galaxies

The Hubble Space Telescope

Scientists need to see far beyond Andromeda in space and time in order to find out more about the big bang. Today's very large telescopes can gather even very dim light from space. Other telescopes gather radio waves and infrared rays from space and use them to form pictures that we can study.

One telescope that has opened a window on the early universe is the Hubble Space Telescope. The Hubble is a telescope that orbits, or circles, Earth high above the atmosphere. The atmosphere is the thin layer of air around Earth. In space, outside of Earth's atmosphere, haze, pollution, and lights do not block Hubble's view of deep space objects.

Pictures taken by the Hubble show galaxies that are farther away than anything we've ever seen before. They lie 12 billion light-years out in space. Seeing these galaxies takes us back nearly to the beginning of the universe itself.

Pioneers of the Big Bang Theory

The big bang theory is the basis for our study of the universe today. It is hard to imagine that the universe sprang out of nowhere. Even if the universe did, however, the big bang theory did not. It is based on clues we find in the universe. The biggest clue is the fact that the universe is still expanding.

One hundred years ago, we did not know about the universe's expansion. Many scientists thought the universe was the same size it had always been, and that it had no starting point.

In the early 1900s, some scientists began wondering about the universe and whether it could be expanding. They used mathematics to explain how this might be true. One of these scientists was named Georges Lemaître.

Georges Lemaître

Lemaître, however, went one step further. In 1931, he said that the universe may have been expanding ever since it began. Perhaps, he suggested, it began as a "cosmic egg." This "egg" exploded and sent the universe expanding in all directions. Because of his ideas, Lemaître is known as "the Father of the Big Bang."

Edwin Hubble at work

It was one thing to show on paper that the universe was expanding. Was there any proof in the sky, however? The first clue came in 1923 when astronomer Edwin Powell Hubble discovered galaxies.

Astronomers had seen these galaxies before. They did not think the galaxies were far-off systems of stars, however. They thought the galaxies were just clouds of dust and gas inside our own Milky Way Galaxy. Such clouds are called nebulae.

Hubble was able to study the nebulae through a powerful telescope. He spied stars inside some of them. Some of these were stars called Cepheid variables. They are very bright at times, then go dim for a while before brightening again. Hubble used the light from these stars to figure out distances in space.

When Hubble did the math, he found that these stars were much, much too far away to be inside our own galaxy. They must be inside other galaxies. That meant there could be many other galaxies. Billions of them!

Cartwheel Galaxy

Then, in 1929, Hubble found that other galaxies are moving away from us. At the same time, our galaxy is moving away from all the other galaxies, too. Not only that, but the galaxies farthest away from us were moving the fastest.

All these galaxies were on the move because the universe was still expanding and carrying them away from each other. Like the raisins in the baking bread.

How could Hubble see that the galaxies were moving away from us, as well as from each other? They are so far away, he would never be able to actually see them move. However, he *could* study the light they gave off.

Visible light that comes from the Sun and other stars is made up of all the colors of the rainbow. Scientists can break up the light coming in from a star to see the rainbow, or spectrum, of colors that form it. These colors are measured in wavelengths. They can be drawn to look like a series of rippled waves, with red light having the longest wavelengths and blue light having the shortest.

Still, something peculiar happens to starlight. The expansion of the universe changes a star's spectrum. The light is stretched out during its long journey. Its wavelength is made longer. This makes its spectrum look more red than if the universe was not expanding. When this happens, scientists say that the light is "red-shifted."

Objects moving toward us, on the other hand, show a blue shift. The wavelengths they give off are squeezed instead of stretched. They get shorter. The spectrums are "blue-shifted." Blue-shifting helps us see the movements of stars that orbit each other.

Galaxies that are very far away show a bigger red-shift than galaxies closer by. The bigger the red-shift is, the farther away a galaxy is, and the faster it is moving away.

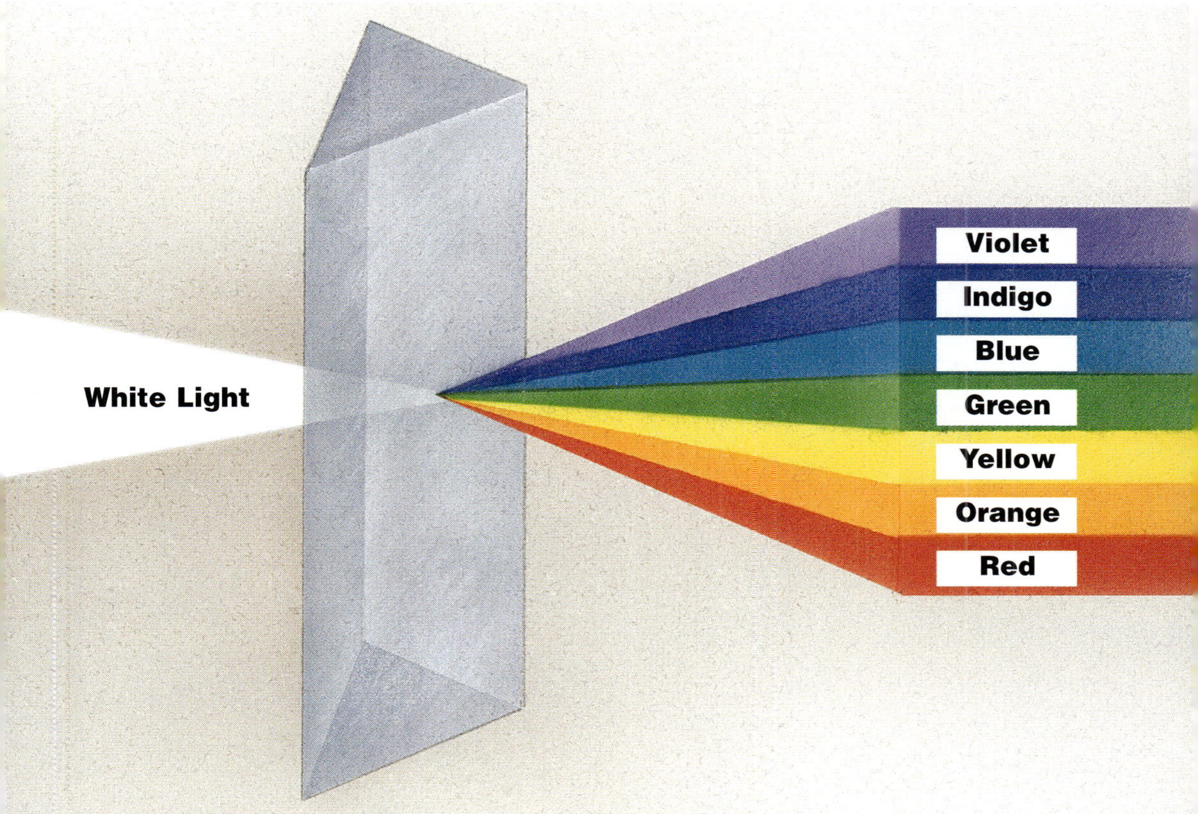

Visible light looks white when it is made of all the colors of the rainbow. A prism separates these colors.

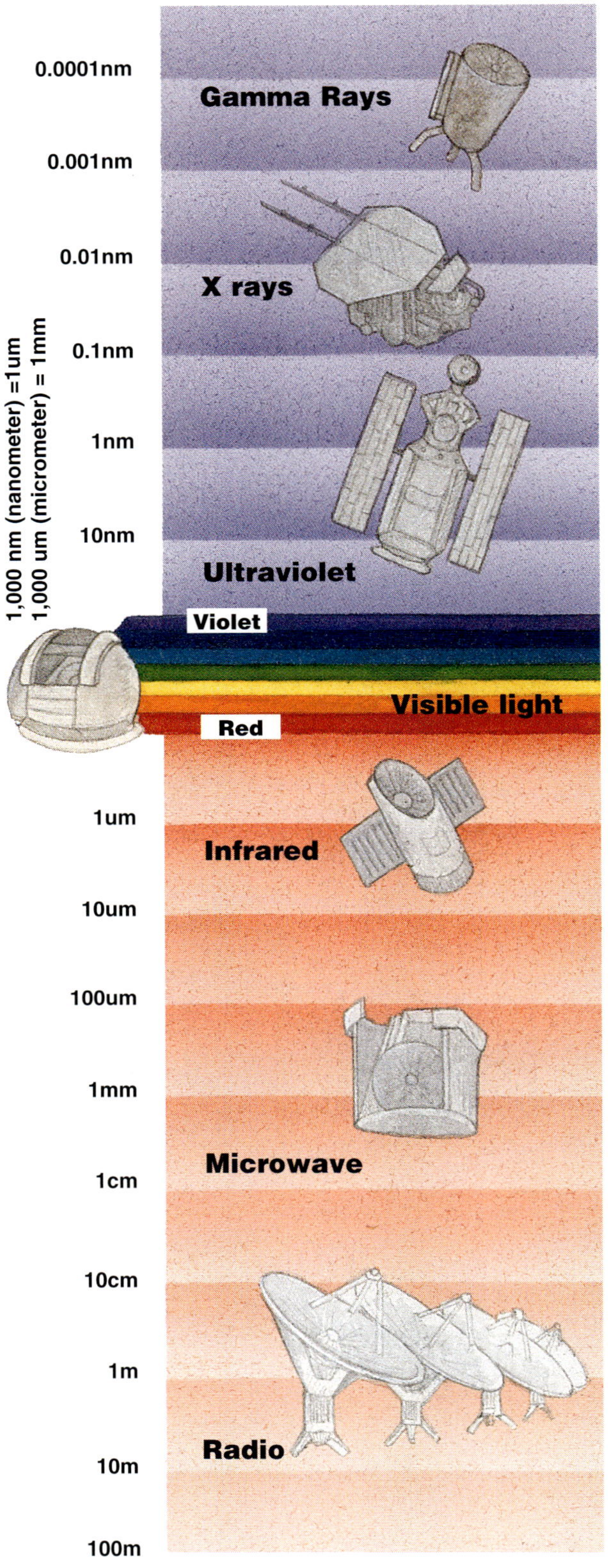

The Electromagnetic Spectrum

Light is also called radiation. Visible light is only a part of a wide band of radiation called the electromagnetic spectrum.

The electromagnetic spectrum includes "invisible" light, such as X rays and ultraviolet rays. These are rays that you cannot see. The spectrum includes infrared rays, radio waves, and microwaves, too. Objects in space give off all these kinds of radiation.

We cannot see these other kinds of radiation. We can only see the visible-light part of the electromagnetic spectrum. We can measure these kinds of radiation with scientific instruments, however. Then we can use that information to tell us about objects in space. That is how another important part of the mystery of the big bang was discovered.

The first hint of this discovery came in 1948. That was when a scientist named George Gamow had an idea about "noise." If the big bang actually happened, Gamow reasoned, then the universe should be filled with an incredible "echo" from the past. This echo would be some kind of radiation that was everywhere, and could be detected with the right instruments. Gamow also knew that the universe had cooled down as it expanded and got older. So he also figured out what temperature this radiation should be.

Arno Penzias and Robert Wilson at work

Fifteen years later, two scientists found Gamow's radiation. Their names were Arno Penzias and Robert Wilson. They were studying the sky using a special kind of telescope called a radio telescope. They were trying to find out where radio waves were coming from in the universe. There seemed to be a lot of "noise" getting in the way. They discovered lots of microwave radiation that did not seem to come from a single object or place.

Microwave radiation has a longer wavelength than visible light. It is on the short-wavelength end of radio waves. It forms a tiny part of the static you can see on a TV set that isn't tuned in properly.

The microwave radiation in space seemed to be part of a "buzz" coming from everywhere.

The scientists measured the temperature of this mysterious radiation. They found it to be about -454° Fahrenheit (-270° Celsius). That is almost four times colder than the coldest spot on Earth, in Antarctica. It is also close to the temperature that George Gamow said it would be. Scientists called it "background radiation." It seemed to be a "left over" from the big bang.

In 1989, the United States launched a satellite to study the background radiation. This satellite was called the Cosmic Background Explorer, or COBE for short. COBE measured the temperature of the background radiation.

COBE found that the temperature of the background radiation was not the same all over. It found some areas that were a little hotter. It found others that were a little colder. The differences in temperature were tiny. They showed that there were "wrinkles" and "bumps" in space.

The Cosmic Background Explorer, COBE

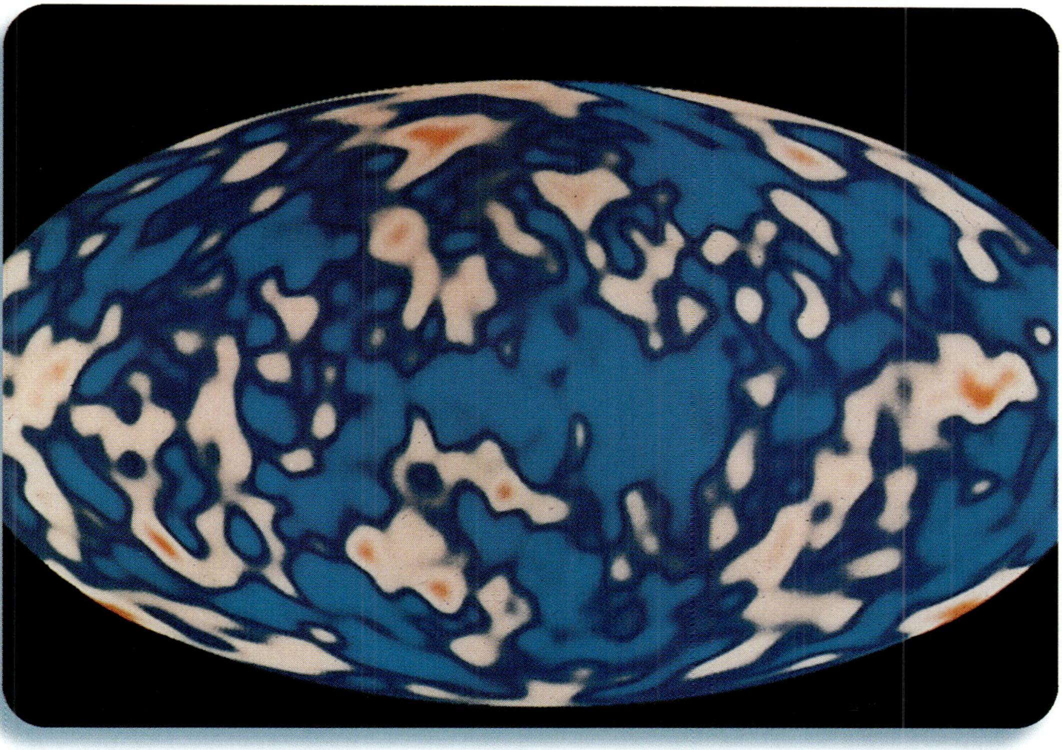

A COBE photograph of background radiation in the universe

This discovery was very important. Before COBE, it seemed that the background radiation was exactly the same everywhere. It was like a smooth sheet of glass, without a bump or a ripple anywhere.

This posed a problem for astronomers. If space were the same everywhere, then why were there galaxies? Why wasn't all the matter in space sprinkled evenly everywhere? Why were stars and other space objects clumped together in different places with empty space in between?

COBE showed that the background radiation was not exactly the same in all places. The slight temperature differences proved this. Seeing space through COBE's "eyes" was like looking closely at a smooth sheet of glass with a microscope. The smooth glass turns out to be full of pits and bumps!

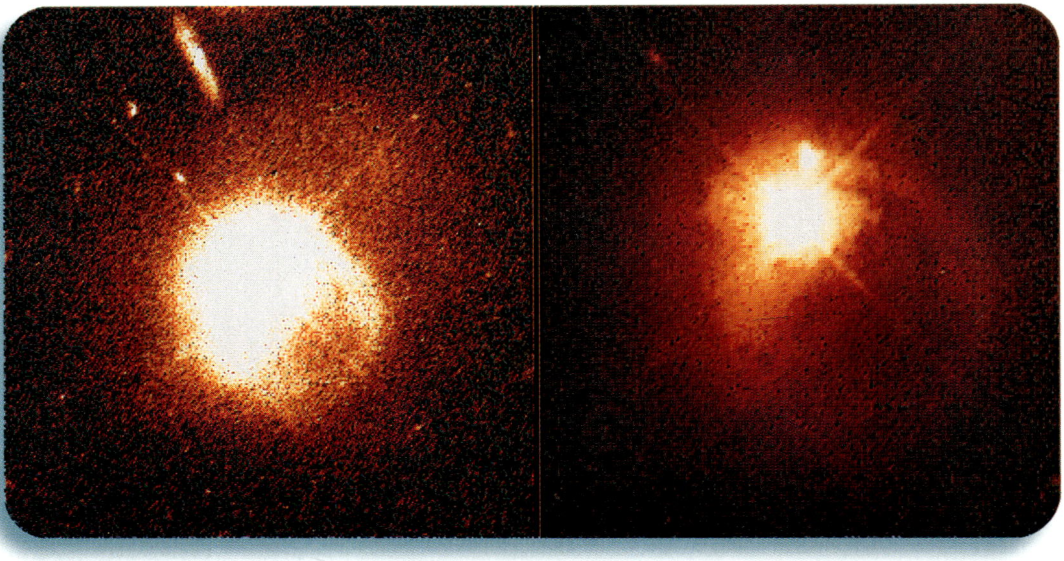

A Quasar

What's next?

This is one of the biggest questions facing cosmologists today. Will the universe expand forever? Will it expand for a while, then slow down more and more without ever coming to a complete halt? Perhaps it will start to shrink and end up in a big crunch—the reverse of the big bang.

To answer these questions about the future, cosmologists continue looking into the past. COBE took scientists nearly 15 billion years back in time. It gave us a glimpse of the universe as it was just half a million years after the big bang. The Hubble Telescope has revealed galaxies born just a few billion years after the big bang. Telescopes have also shown us quasars—bright objects that glow at the very edge of the universe and therefore at the edge of time.

What this technology has not revealed is the "missing mass" of the universe—and it is this "missing mass" that may hold the key to the universe's future.

Cosmic gas clouds of matter taken by the Hubble Space Telescope

The term "missing mass" refers to all the material in the universe that we cannot see. It is also called "dark matter." Scientists think that as much as 99 percent of the material in the universe is made up of dark matter. This dark matter does not give off light or any other radiation we can detect.

How can scientists know that dark matter exists if they cannot see it? One clue is the effect of dark matter's gravity. Gravity is the force that makes objects attract each other. It is what keeps you on the ground, makes the Moon orbit Earth, and Earth orbit the Sun. All matter has gravity—from the tiniest atom to the biggest galaxy.

Scientists have found that we simply cannot detect enough matter in the universe to explain all the gravity that exists. There isn't even enough matter to produce enough gravity to hold galaxies together. However, the galaxies have not fallen apart.

The extra gravity must belong to matter that is not like ordinary matter. That matter is dark matter.

Scientists do not yet know exactly what forms dark matter. They have some ideas, however. Some dark matter may be very large, such as black holes, dead stars, and asteroids. Much dark matter, however, might be very tiny.

One type of tiny particle that scientists are studying closely is called a neutrino. A neutrino is smaller than an atom. The Sun and other stars give off trillions of neutrinos. Neutrinos are very hard to detect, however. They may form a large part of dark matter.

Whatever dark matter is, it seems to have helped to shape the first galaxies with the power of its gravity. It remains to be seen whether dark matter's gravity is enough to stop the expansion of the universe, however, let alone reverse it and cause a big crunch!

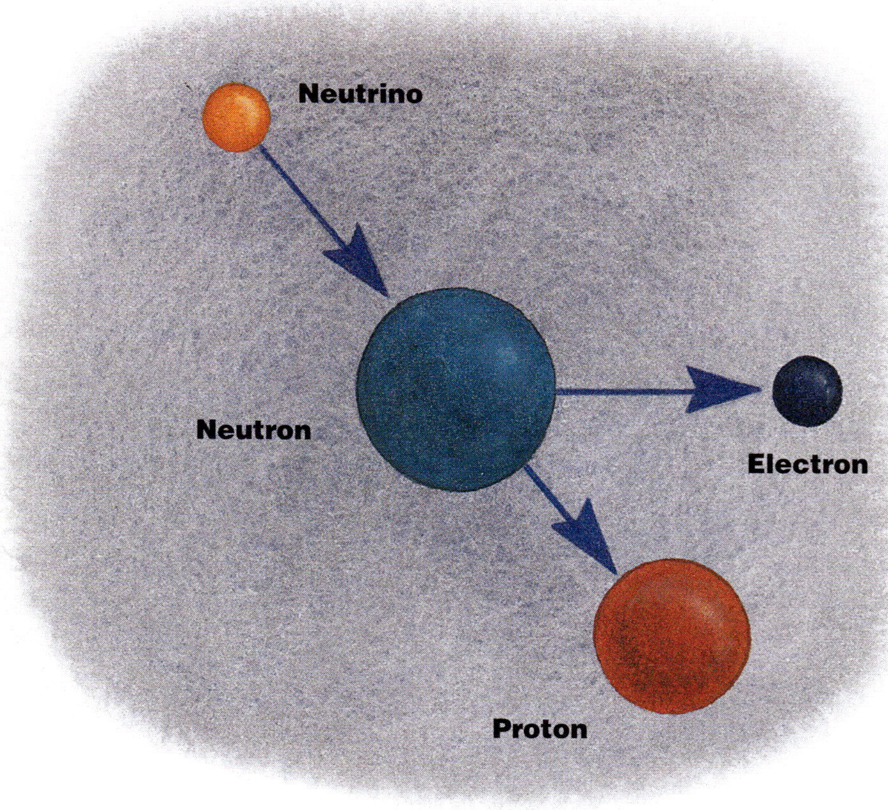

A neutron changes into an electron and proton.

Diagram of Inflationary Expansion of the Universe

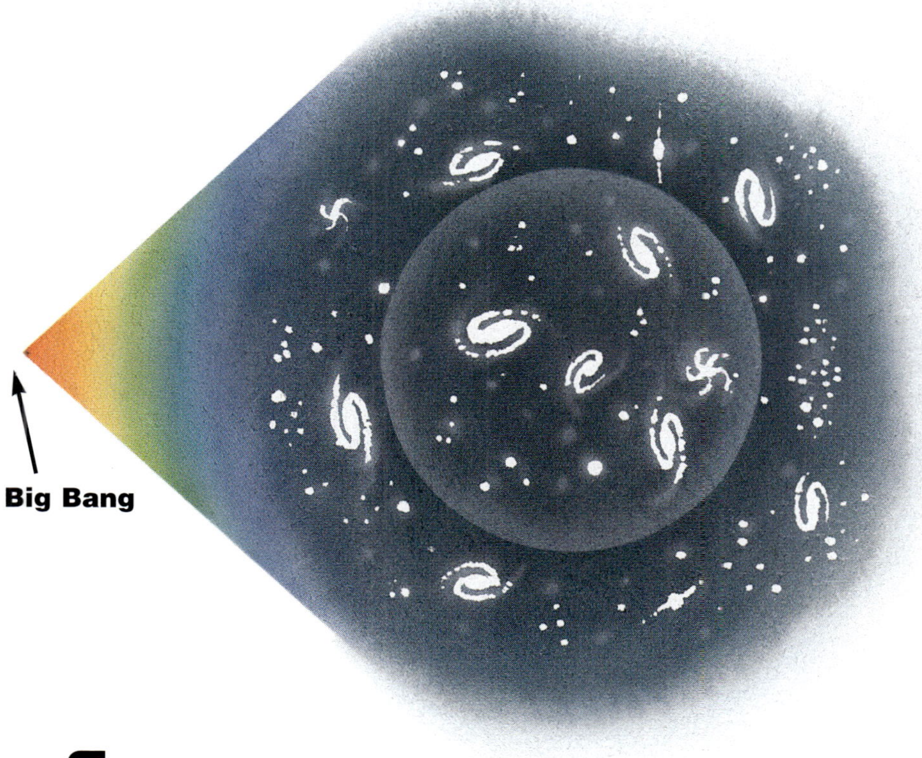

Cosmologists have also been given food for thought by a scientist named Alan Guth. Guth worked out that the early universe must have expanded much more quickly than it is expanding today. He stated that the universe was smaller than a proton at first. Then it grew to be trillions of times bigger in the merest fraction of a second.

The result of this fast growth, or "inflation," was not only the universe we can see, but also a gigantic portion of the universe beyond it that we cannot see. This idea is called the "inflationary theory."

The inflationary theory helps explain why one part of the universe is so similar to another part, even though those two parts were never in contact with each other. The "inflationary" universe is so incredibly big that it makes the galaxies and other "lumps" in the universe that we can see seem very small.

At this time, many scientists agree that the universe will continue to expand. In 1998, some scientists even suggested that the universe's expansion is speeding up! They came up with this idea after studying supernovas using the Hubble and other big telescopes. The supernovas were up to 7 billion light-years away from Earth. They were clearly moving away from each other.

These scientists wonder if there is a force at work in the universe that is the opposite of gravity. It is called "repulsive gravity." Such a force would make galaxies repel, or push away, other galaxies.

The universe's future may be less of a mystery than its prehistory—that is, what came before the big bang. We have no way of finding out. There may have been nothing—no time, no space, no matter. There may have been another universe. Our universe may even be just one of many universes, created by one of many big bangs.

We may never know the answer. However, we can continue using technology to sweep the skies for clues. We can gaze up at the stars and wonder at the origin of it all.

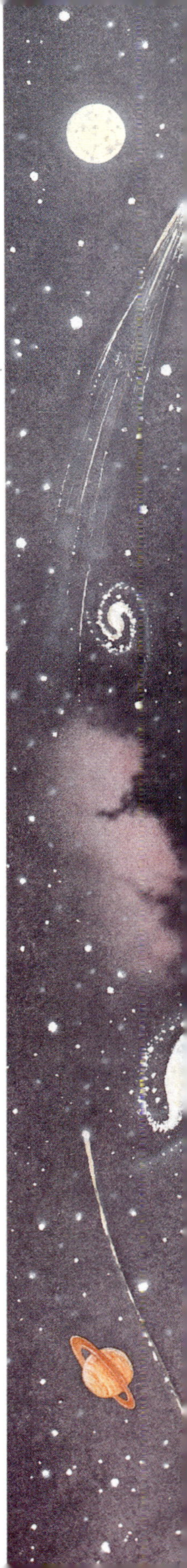